iScience
Readers

Balance and Motion:
Toying with Gravity

by Emily Sohn and Joseph Brennan

Chief Content Consultant
Edward Rock
Associate Executive Director, National Science Teachers Association

NORWOOD HOUSE PRESS
Chicago, IL

Norwood House Press
PO Box 316598
Chicago, IL 60631

For information regarding Norwood House Press, please visit our website at
www.norwoodhousepress.com or call 866-565-2900.

Special thanks to: Amanda Jones, Amy Karasick, Alanna Mertens, Terrence Young, Jr.

Editors: Jessica McCulloch, Barbara J. Foster, Diane Hinckley and Kasi Valjat
Designer: Daniel M. Greene
Production Management: Victory Productions, Inc.

Paperback ISBN: 978-1-60357-303-0

The Library of Congress has cataloged the original hardcover edition with the following
call number: 2011016655

Manufactured in the United States of America in North Mankato, Minnesota.
296R—082016

Contents

Note to Caregivers:

Throughout this book, many questions are posed to the reader. Some are open-ended and ask what the reader thinks. Discuss these questions with your child and guide him or her in thinking through the possible answers and outcomes. There are also questions posed which have a specific answer. Encourage your child to read through the text to determine the correct answer. Most importantly, encourage answers grounded in reality while also allowing imaginations to soar. Information to help support you as you share the book with your child is provided in the back in the **Additional Notes** section.

Words that are **bolded** are defined in the glossary in the back of the book.

Don't Fall!

Acrobats do lots of tricks. They walk across skinny ropes. They stand on their hands. How do they keep their **balance**?

Read on. You'll learn about balance and **motion.** Maybe you'll pick up some tricks of your own!

Make a Toy Acrobat Balance

You just came from the circus. The acrobats amazed you! How did they keep their balance? You decide to make a toy acrobat to find out.

Materials

- 8 craft sticks
- small foam ball (about the size of a golf ball)
- modeling clay
- empty 4-ounce yogurt container
- craft glue
- sandpaper
- scissors

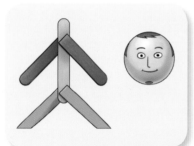

Cut 1 inch (2.5 centimeters) off the end of a craft stick. The larger piece is the body. Cut two sticks for arms and legs.

Glue the pieces together to make a stick person. Use sandpaper to make the feet pointy. Use the foam ball as the head.

Your acrobat needs to balance on the yogurt cup. But the toy is too tippy. It keeps falling over. A pole might help.

How should you build the pole?

Idea 1: Make a straight pole. Glue two sticks together to make a long pole. Put clay weights on the ends to balance it.

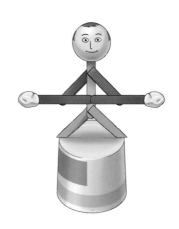

Idea 2: Make a pole bent upside down. Glue three sticks in an upside-down U shape. Clay weights will hang down below the acrobat's feet.

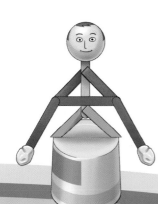

Idea 3: Make a pole bent right-side up. Glue three sticks in a right-side up U shape. Clay weights will be over the acrobat's head.

Which way do you think will work best?

Magic Forks

Materials

• 2 matching dinner forks
• toothpick
• tall glass of water

You can balance two forks on a toothpick.

All you need is a glass of water.

toothpick

glass of water

forks

The pointy parts of forks are called tines. Stick the tines of one fork through the tines of the other. The forks will make a V. You will see little squares between the tines.

Put an end of the toothpick up through one of the squares. It should stay in place.

Put the other end of the toothpick on the edge of the glass.

Let go.

The forks will hang in mid-air!

What Is Gravity?

When you fall, you don't fall up. You fall down. **Gravity** explains why. This is a force that pulls on objects. All objects are affected. Big, heavy objects make more gravity force than small, light objects. Gravity force from Earth has a greater effect on you than your gravity force has on Earth. That's why you don't float into space. The center of Earth pulls you in.

How does gravity affect your toy acrobat?

Gravity is the reason we fall down, not up.

What Is Balance?

Ballerinas have very good balance.

You do a lot all day. You walk, run, and sit. You climb, read and sleep. All of these actions use balance.

Most of the time, you are **stable.** This means you do not fall down. You are balanced even when you are moving.

These children are walking on a fallen tree. Are they stable when they are moving? Do you think they will fall?

Some things are **unstable.** They are not balanced. Gravity makes them fall.

Look at your acrobat. Is it stable or unstable?

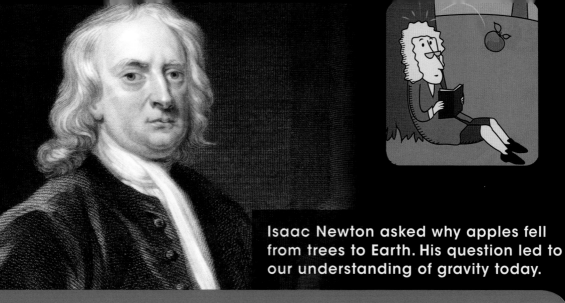

Isaac Newton asked why apples fell from trees to Earth. His question led to our understanding of gravity today.

Connecting to History

Sir Isaac Newton

Isaac Newton lived from 1642 to 1727. One day, he saw an apple fall from a tree. He wondered why it fell down and not up, or sideways. He figured out that a force coming from Earth was pulling on the apple. This pulling force was gravity.

Then Newton thought, "If Earth pulls on apples, maybe it pulls on the Moon, too. Maybe that's why the Moon stays with Earth in space." He was right.

Now we know that gravity affects all objects on Earth and in space.

What Is the Center of Gravity?

Look at a seesaw. It balances if both riders are the same size. The seesaw tips over a spot in the middle. This is the pivot point. If the seesaw is balanced, the pivot point is the **center of gravity.**

pivot point

A low center of gravity helps with balance. That's why surfers crouch down. They are more likely to fall if they stand up straight.

Are you usually more stable on one foot or two?

Think about your acrobat. Where is its center of gravity? How could you make this point lower?

This seesaw is not balanced. The center of gravity is not at the pivot point.

Which fruit is heavier?

A **balance** is like a seesaw. It can show you if two things weigh the same. If they do, they stay at the same height. If they have different weights, the heavier side drops lower.

You cannot change the weights of the fruits in this picture. But you can make them balance with each other. You just need to move the center of gravity. What is a way to do this?

Gymnasts

Gymnasts are athletes. They are also performers. They jump. They flip and twist. They hang from rings. And they balance on beams.

Gymnasts practice a lot. They have to learn to control their centers of gravity. A lot of eyes are watching!

These gymnasts have learned to work with gravity. That's how they balance so well. (They are also very strong!)

For how long can you balance on one foot?

Try this! Balance on one foot and hold your hands close to your chest. Now stretch your arms wide. Which way feels more stable? Keep one foot in the air. Now put something heavy in one hand. Then put heavy things in both hands. How do you have to move to stay balanced? Does this activity give you any ideas that you can use to solve the iScience Puzzle?

This man is performing with 20 hoops. How many can you spin at a time?

Did You Know?

A plastic toy hoop is made to spin. It moves around your hips.

The hoop's center of gravity is at the center of the hoop.

The hoop's center of gravity moves around your body as you move.

We tend to lean forward when we run. That keeps our center of gravity lower than it is when we stand up straight. Leaning makes us more stable.

Your center of gravity changes in relation to Earth as you move. Think about what your body does when you run. You likely lean forward.

If you didn't, you might fall over.

Why might it be hard to run on a tightrope?

Which way makes the toy acrobat balance?

Idea 1: Make a straight pole.

Idea 2: Make a pole bent upside down.

Idea 3: Make a pole bent right-side up.

Idea 2 will work best. You want the center of gravity to be very low. The hanging weights will lower the center of gravity of the acrobat. The toy will balance more easily.

How does the boy move his arms to help him balance on the pole?

Play with balance and motion.

Lay a broom across your hands. Spread your hands far apart. Is it hard to balance the broom? Now, move your hands so they are touching. What do you have to do to balance the broom? Can you balance it on one hand or finger? What if you try to move and balance the broom at the same time?

Try the same thing with other objects. Which are hardest to balance? Which are easiest? What did you learn about balance and motion?

What role does gravity play in balance and motion?

Glossary

acrobats: gymnasts who perform before an audience.

balance: 1. a stable position in which a person or an object does not tip one way or the other. 2. a device that shows how much one object weighs compared to another object.

center of gravity: the point in an object around which its weight balances.

gravity: a pulling force between objects.

gymnasts: people who can hang or pull themselves up by rings or bars, and walk on a thin board.

motion: movement.

stable: not likely to fall or go off balance.

unstable: not likely to stay the way it is.

Further Reading

Cool Gravity Activities: Fun Science Projects about Balance (Cool Science), by James Hopwood. Checkerboard Books, 2007.

The Mystery of Gravity (Story of Science), by Barry Parker. Benchmark Books, 2008.

Escapade Direct, Play With Gravity, Activities for Children.
http://www.escapadedirect.com/plwigr.html

TLC, How Stuff Works, Family.
http://tlc.howstuffworks.com/family/science-experiments-for-kids9.htm

Additional Notes

The page references below provide answers to questions asked throughout the book. Questions whose answers will vary are not addressed.

Page 10: Gravity pulls on the toy, helping it balance on the cup.

Page 14: You are more stable on two feet. Your acrobat's center of gravity is low, near the top of its legs. To lower its center of gravity, you could make its arms longer or use heavier weights (more clay).

Page 15: You can add more weight to the apple side or move the apple closer to the orange. Caption question: The orange is heavier.

Page 19: Running on a tightrope would make the rope move with every step you took. Every time the tightrope moved, your body would move. That would make your center of gravity change. You would have a hard time finding and keeping your center of gravity.

Page 20: Caption question: The boy moves his arms and keeps them low so he can try to find his center of gravity.

Index